# Space Coloring Book

**Stress Relieving Space Designs For Anger Release, Relaxation And Meditation, For Kids, Teens And Adults**

**- JESSICA PARKS -**

The pages of this book are suitable for crayons and colored pencils. Also suitable for framing.

To prevent bleed through when coloring, scroll to the end of the book and rip the "tear me" page.

ISBN-13: 9781730772788

© Copyright 2018

Jessica Parks / BlissRockHub OU

ALL RIGHTS RESERVED.

No part of this publication may be reproduced or transmitted in any form whatsoever, electronic, or mechanical, including photocopying, recording, or by any informational storage or retrieval system without express written, dated and signed permission from the author/publisher.

# Tear me...

Use this page

to prevent

bleed through

when coloring